PROGRESS IN

WORKBOOK 3

GEOGRAPHY

UNITS 11–15

KEY STAGE 3

DAVID GARDNER

HODDER
EDUCATION

LEARN MORE

NAME: ....

CLASS: ....

The Publishers would like to thank the following for permission to reproduce copyright material.

Photo credits

p. 8 © Marie Tharp, Bruce Heezen and Heinrich Berann / Lamont Doherty Earth Observatory; **p. 14** © David Gardner; **p. 30 top left** © robertharding / Alamy Stock Photo; **p. 30 top right** © Dave Ellison / Alamy Stock Photo; **p. 30 bottom** © David Gardner; **p. 33** © David Gardner.

Acknowledgements

All OS maps used throughout this book have been reproduced from Ordnance Survey mapping with permission of the Controller of HMSO. © Crown copyright and/or database right. All rights reserved. Licence number 10003470.

Ordnance Survey (OS) is the national mapping agency for Great Britain, and a world-leading geospatial data technology organisation. As a reliable partner to government, business and citizens across Britain and the world, OS helps its customers in virtually all sectors improve quality of life.

Orders: please contact Hachette UK Distribution, Hely Hutchinson Centre, Milton Road, Didcot, Oxfordshire, OX11 7HH. Telephone: +44 (0)1235 827827. Email education@hachette.co.uk Lines are open from 9 a.m. to 5 p.m., Monday to Friday. You can also order through our website: www.hoddereducation.co.uk

ISBN: 9781510442986

© David Gardner 2019

First published in 2019 by

Hodder Education,

An Hachette UK Company

Carmelite House

50 Victoria Embankment

London EC4Y 0DZ

www.hoddereducation.co.uk

Impression number   10 9 8 7 6

Year   2023

Cover photo © arquiplay77 – stock.adobe.com

Illustrations by Aptara Inc.

Typeset in India by Aptara Inc.

Printed in the UK

A catalogue record for this title is available from the British Library.

Contents

COMPLETED

Contents

Hodder & Stoughton Limited © David Gardner 2019

COMPLETED

15 Climate change and the Earth's future

Student's Book
pages 202–203

11.1

11.1 Can we ever know enough about earthquakes and volcanoes to live safely?

Creating fact files for a volcanic eruption and earthquake

Fact file: Volcan de Fuego, Guatemala

Location: Mark the location on the map and add latitude and longitude

Date of eruption:

_____

Impact:

_____

_____

Complete after Lesson 11.5

State the type of plate boundary this volcano occurred at, giving evidence to support your choice:

_____

_____

Fact file: Van, Turkey

Location: Mark the location on the map and add latitude and longitude

Date of earthquake:

_____

Impact:

_____

_____

Complete after Lesson 11.5

State the type of plate boundary this earthquake occurred at, giving evidence to support your choice:

_____

_____

11.2 Do continents fit together like jigsaw pieces?

Alfred Wegener's theory of continental drift

1 Explain in a sentence Alfred Wegener's theory of continental drift.

_____

_____

_____

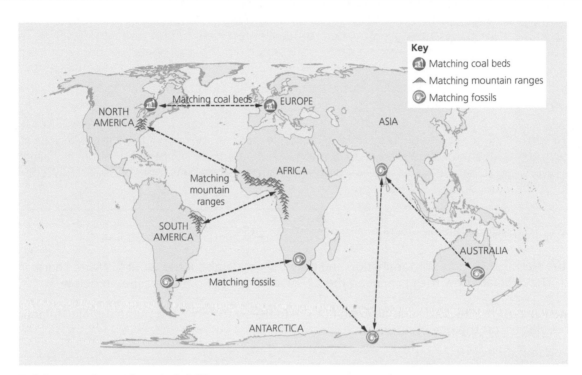

Evidence of continental drift

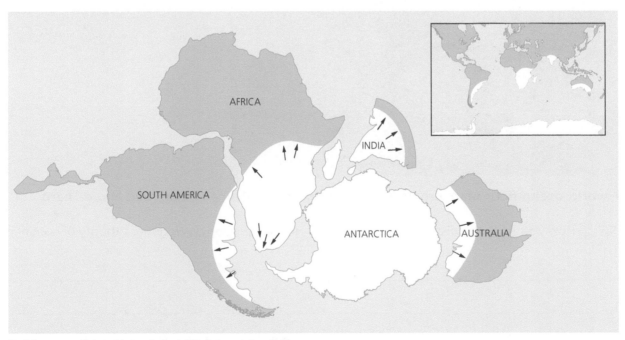

Evidence of continental drift from glaciation

11.3

Student's Book
pages 206–207

11.3 Where are the world's earthquakes, volcanoes and mountain belts?

Tharp and Heezen's discoveries about the ocean floor

More than 70 per cent of the Earth is made up of oceans, which are the least known areas of the planet.

Marie Tharp was a geologist and cartographer from the USA. She worked in partnership with Bruce Heezen. Tharp was the first person to scientifically map the ocean floor, publishing the world ocean floor map in 1977.

Up until this time, the ocean floor was believed to be a flat plain of mud. Tharp and Heezen wanted to map the ocean floor in order to understand its geology. They collaborated for many decades, from the 1950s into the 1970s, gathering information. Bruce Heezen went out on research vessels to sea and collected data. Much of the raw data came from soundings, or sonar measurements, of the ocean depths.

Tharp and Heezen discovered that the ocean floor was not flat but covered with various kinds of geological features like canyons, ridges, and mountains, just like on the Earth's above-ground continents.

In 1953, Tharp made a remarkable discovery – the Mid-Atlantic Ridge, a chain of mountainous volcanoes that runs north to south through the ocean. She observed a depression in the ridge that appeared to be a continuous crack along its length. Her discoveries provided new evidence supporting Wegener's theory of continental drift, or seafloor spreading.

1 Go to https://news.nationalgeographic.com/2017/02/marie-tharp-map-ocean-floor/.

Watch the National Geographic animation and read the article about the work of Marie Tharp and Bruce Heezen. They explain how they produced the 1977 world ocean floor map, shown below.

2 Annotate the copy of the 1977 world ocean floor map below to show and name the Mid-Atlantic Ridge, and other ridges, on the ocean floor.

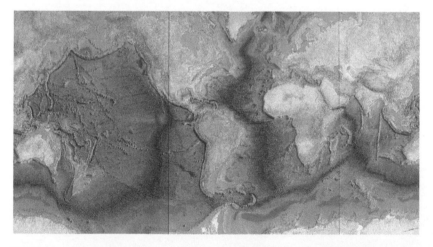

1977 world ocean floor map, created by oceanographers Bruce Heezen and Marie Tharp

3 Explain how you think the discoveries of Tharp and Heezen supported Wegener's theory of continental drift.

_____

_____

_____

The theory of plate tectonics

1 Watch the National Geographic online video about the theory of plate tectonics:
www.nationalgeographic.org/media/plate-tectonics/.

2 Complete the following paragraph about the Earth's structure and the theory of plate tectonics using the words from the box below.

volcanoes	outer core	continental plates	lithosphere	earthquakes	inner core	lava
crust	plates	hottest	mantle	plate boundaries	oceanic plates	magma

The Earth is made up of several layers. The outer layer is one of the Earth's spheres; it is called the
_____. This layer has a thin layer of rock on top; it is called the _____. This
layer is 5–30 km thick. The next layer is much thicker; it is called the _____. It is made of
liquid rock called _____. Where the earth's outer layer is thin, this liquid rock can explode
onto the surface as _____, forming _____. The next layer is called the
_____; here material is in a liquid state. The layer at the centre of the Earth is called the
_____. This is the _____ part of the Earth.

The outer layer of the Earth is made up of slabs that sit on top of the mantle; they are called
_____. There are two types: _____, which are 50–100 km thick, and
_____, which are thicker – up to 200 km thick. These slabs of crust are constantly moving
and meet in various ways along their edges at _____. This is where most volcanoes,
_____ and mountain belts occur.

3 Label the key features of the Earth on the diagram below, including: lithosphere, continental crust, oceanic crust, mantle, outer core, inner core. Add labels to show which layers of the Earth are solid and which are liquid.

The layers of the Earth

4 Write two sentences to explain why you think the slabs of plates move.

_____

_____

_____

_____

_____

11.5 What happens at plate boundaries?

Types of plate boundary

The diagrams below show the four types of plate boundary.

1 Add a title in the space provided on each diagram naming the type of plate boundary.

2 Annotate each diagram to explain what is happening at each boundary and what features can be found there.

3 Compare the diagrams with a map showing the world's tectonic plates. Identify places in the world where each plate boundary type occurs. Name the plate names at these boundaries on the diagrams.

Type of plate boundary _____

Type of plate boundary _____

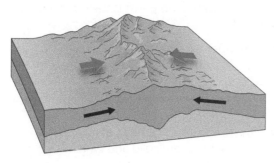

Convergence where two continental lithospheric plates meet

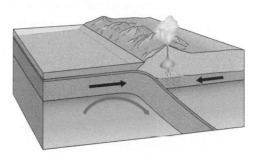

Convergence at the boundary of an oceanic lithospheric plate and a continental lithospheric plate

Type of plate boundary _____

Type of plate boundary _____

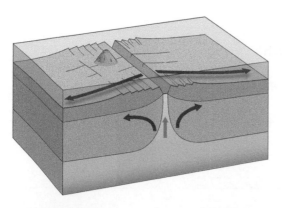

Convergence where two oceanic lithospheric plates meet

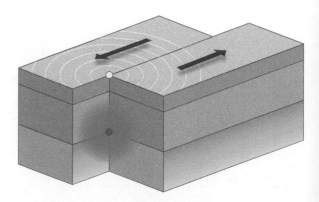

Conservative or transform plate boundaries

11.6 What do we know about earthquakes?

Researching about earthquakes

1 Match the correct endings to the beginnings of these sentences.

Beginnings	Endings
An earthquake is	is called the epicentre.
The location below the Earth's surface where the earthquakes start	waves of energy caused by the sudden movement of the plates.
The area on the surface directly above the focus	along fault lines.
Seismic waves are	a sudden violent movement of the Earth's surface.
When an earthquake occurs, plates slide	mainshock; it is always followed by after shocks.
The largest main earthquake is called the	reduces the further away it is from the focus.
The power of the earthquake	is called the focus.

2 Use the internet to search for and investigate a recent earthquake. You will find the following websites a useful start:

- USGS Earthquakes https://earthquake.usgs.gov/earthquakes/
- BBC News www.bbc.co.uk/news

You could also use a search engine such as Google to search for major earthquakes this year.

3 Complete the fact file for your chosen earthquake.

> ### Earthquake fact file
> Location of earthquake: _____
>
> Date of earthquake: _____
>
> Focus of earthquake: _____   Country latitude and longitude:
>
> Epicentre: _____   _____
>
> Type of plate boundary: _____
>
> Names of plates: _____
>
> Impact of earthquake on people: _____
>
> _____
>
> _____
>
> _____
>
> **4** Compare this earthquake with the Nepal earthquake of 2015. Identify the similarities and differences.
>
> **Similarities**   **Differences**
>
> _____   _____
>
> _____   _____
>
> _____   _____

11.7 Can people manage risk living in earthquake zones?

Create a public information poster preparing people for an earthquake

Imagine you have been asked to produce a poster for the government of a country susceptible to earthquakes, providing guidance on how people should prepare for a quake.

Use the ideas below to create your poster, either drawing the poster by hand or using a computer to produce the final version. Stick your final poster in the space below over the guidance.

Stick your poster here.

Carry out some internet research to find posters that have already been produced about guidance on preparing for earthquakes to help you gather some ideas for your own poster. You can also think about the posters that you investigated in Lesson 11.7.

The governments of many countries produce guidance for their populations in order to save lives. The following websites show real guidance that is provided for people as a public service and can be used to give you some ideas for your poster.

● USA Department of Homeland Security www.ready.gov/earthquakes
● American Red Cross www.redcross.org/get-help/how-to-prepare-for-emergencies/types-of-emergencies/earthquake.html
● New Zealand government http://getthru.govt.nz/disasters/earthquake/

You could perform your own research on other countries to find more guidance.

Things to consider when designing your poster:
● The name of the country.
● A catchy title to hook people in, for example, 'Drop, cover, hold on'.
● The use of colour in the design.
● The size and type of font.
● Photos and diagrams to provide the guidance; for example, you could design a clipboard checklist of things for people to do before, during and after an earthquake.

11.8 What do we know about volcanoes?

What are volcanoes?

1 Complete the paragraph about volcanoes with the words from the box below.

| lava   hot ash   dormant volcano   poisonous   igneous rocks   active volcano |
| crack in the lithosphere   volcanic bombs   successive eruptions   extinct volcano |

Volcanoes are formed where there is a _____. The

magma can erupt in a number of different forms. _____ flows from the vent or

crack. _____ are lumps of molten rock that solidify as they explode out of the

vent of the volcano. _____ is thrown out of the volcano into the atmosphere and

eventually falls back to Earth. Steam and gas can also be released by volcanoes which is _____.

Volcanoes can build up into mountains with steep sides, as a result of _____

over prolonged periods of time. These steep slopes are made of lava which solidifies into

_____. Volcanoes are found in three states: an _____ is erupting

or has erupted recently, and is likely to erupt again; _____ – this has not erupted for

10,000 years, but could erupt again in the future; and _____ – this hasn't erupted

for the last million years and will probably never erupt again.

Features of a volcano

2 Label the diagram with the following features of volcanoes: crater, main vent, magma chamber, secondary cone.

3 Add three other volcano features onto the diagram.

4 Add a title naming the type of volcano shown in the diagram.

5 Write two sentences to explain how volcanoes form.

_____

_____

_____

_____

11.9 Can people manage risk living near volcanoes?

Mount Etna – a case study of a volcano

1 Study the photo below. The top right-hand corner of the photo leads to the main crater of the volcano. Identify and label the chairlift that leads to the summit. Also label the following features that show how the volcano is used for tourism: restaurants, car and coach parks and tourists.

The upper slopes of Mount Etna in October 2018

2 The volcano erupted in 2017, damaging the chair lift and some buildings. Look carefully at the photo, then draw and label where you think this recent lava flow occurred.

3 Mount Etna has erupted since this photograph was taken. Look at the following online news reports to find out what happened:

- www.bbc.co.uk/news/world-europe-46675110

- www.theguardian.com/world/2018/dec/26/mount-etna-magnitude-earthquake-italy-sicily

4 When did the volcano erupt and what impact did this have on the people of Sicily?

_____

_____

_____

_____

_____

_____

11.10 Can we ever know enough about earthquakes and volcanoes to live safely? Review

11.10

Student's Book
pages 220–221

Review

1 Think about the vision statement for *Progress in Geography* on Flap A of the textbook. Which aspects of becoming a good geographer have you made progress in through your studies of this unit? Write them in the space below, and add evidence about which lessons you made progress in.

_____

_____

_____

_____

_____

_____

_____

_____

_____

_____

_____

_____

_____

_____

_____

_____

2 Label all the places you have studied in this unit on the map of the world.

12.1 What are the challenges and opportunities facing Africa?

What do I think I know about the continent of Africa?

1 Write down five main things you already know about the continent of Africa in the box below.

1

2

3

4

5

2 Watch the Chimamanda Ngozi Adichie TED talk at:
www.ted.com/talks/chimamanda_adichie_the_danger_of_a_single_story.

Write three sentences to explain what she means by a single story of Africa.

_____

_____

_____

3 Many people in the world have a stereotypical view of Africa. Explain why you think this is the case.

_____

_____

_____

_____

_____

4 Look back at your answer to Question 1. Identify one thing you thought you already knew about Africa that might be a stereotypical view, and explain why you think it could be.

_____

_____

_____

_____

_____

The physical landscape of Africa

1 Using a physical map of Africa that you have found in an atlas or on the internet, mark and name:

- the seas and oceans that surround the continent
- what you think are the major physical regions, including mountains, deserts, lakes and basins
- the following major rivers – Nile, Niger, Zambezi.

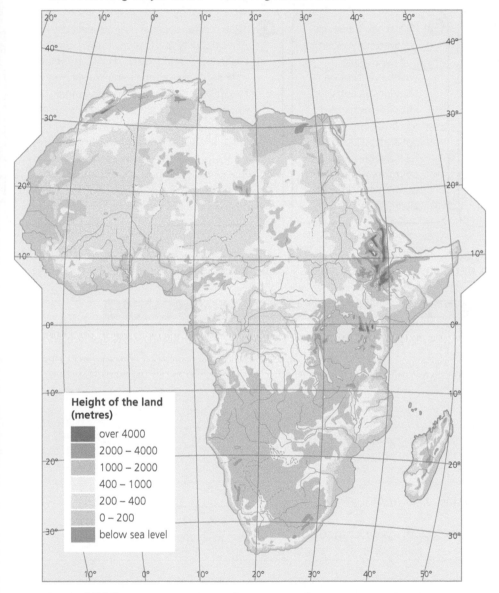

Height of the land (metres)
- over 4000
- 2000 – 4000
- 1000 – 2000
- 400 – 1000
- 200 – 400
- 0 – 200
- below sea level

Map of Africa

2 Write three sentences to describe the physical geography of the continent of Africa.

_____

_____

_____

_____

_____

12.3 How has Africa's past shaped its present?

Different views about the colonialism of Africa

1 Read the three quotes below, taken from the following people:

- Biyi Bandele, a Nigerian playwright living in London: http://news.bbc.co.uk/1/hi/world/africa/4653125.stm
- Cecil Rhodes, British businessman and South African politician, *Confession of Faith*, 1877
- Kwame Nkrumah, Ghana's first Prime Minister, *Africa must Unite*, 1963

A 'I contend that we [Britons] are the first race in the world, and the more of the world we inhabit, the better it is for the human race . . . It is our duty to seize every opportunity of acquiring more territory . . . more territory simply means more of the Anglo-Saxon race, more of the best, the most human, most honourable race the world possesses.'

B 'They took our lands, our lives, our resources, and our dignity. Without exception, they left us nothing but our resentment, and later, our determination to be free and rise once more to the level of men and women who walk with their heads held high.'

C 'Rich former colonial powers like to regard Africans only as victims. Initiatives like Live Aid in the 1980s, led by Bob Geldof, are attempts by well-meaning Westerners to fix Africa's problems by raising money and awareness. They are simply perpetuating the dependency culture created by colonialism. I think the priority now is for Africans to look forward, not back. Africa does need the world's help. But Africa's destiny can be changed for the better only by Africans themselves.'

Match the quotes to the correct people, writing your answers in the boxes below. Give evidence from the quotes to support your choice.

Quote	Source of quote	Evidence for your choice
A		
B		
C		

2 Which quote provides an imperialist view of Africa? _____

Highlight evidence in the quote to support your choice.

_____

_____

3 Quote C refers to a dependency culture created by colonialism. Explain what this means.

_____

_____

4 What does Quote C suggest as a positive way forward for the future of the continent?

_____

_____

12.4 How developed are African countries?

Levels of development in Africa

1 Read the news article at the following website: www.bbc.co.uk/news/uk-20842827.

What did Oxfam discover in its survey of 2000 UK citizens' views about Africa?

_____

_____

_____

2 Go to the Dollar Street website: www.gapminder.org/dollar-street/.

In 'the World' dropdown menu, filter your search to select all the families that live in African countries.

Choose two families at opposite ends of the Dollar Street – one from the poorest and one from the richest. Summarise the information in the table below to compare the two families.

	Poor family	Rich family
Name of the family		
Country		
Income		
How many people in the family?		
Describe the house and its facilities including shelter, furniture, possessions, toilet, heating		
Describe access to food and water		
How do the family earn a living?		
What are their dreams for the future?		

3 What have you learnt about the levels of development for these families in Africa?

_____

_____

4 What are the dangers of a single story view of Africa, as explained by Chimamanda Ngozi Adichie in Lesson 12.1?

_____

_____

_____

12.5 What is the pattern of climate and biomes in Africa?

The climate of Africa

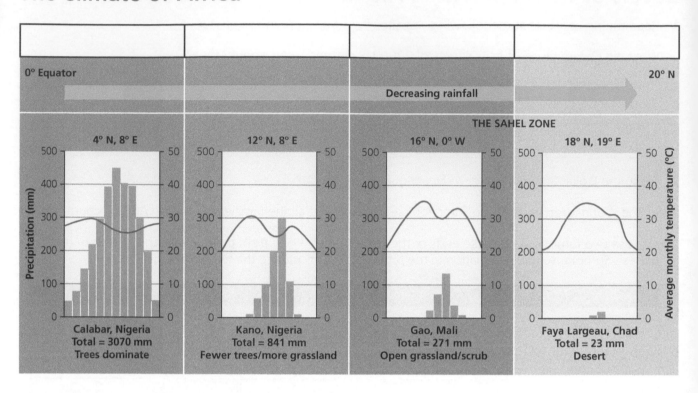

Climate graphs for different African countries

1 Conduct a search on the internet and find an atlas style map of Africa that shows biomes. Compare the four climate graphs shown above with the map.

2 Look for evidence on each graph and the map to decide which biome each climate graph is in – rainforest, deciduous woodland, savannah, desert – and write the name of the biome in the boxes provided above each climate graph.

3 Annotate each climate graph to show two pieces of evidence for each that determined your choices of biomes in Question 2.

4 On the diagram below, label the four biomes that are found in Africa. Annotate the diagram to explain how the circulation of air masses over Africa leads to the formation of the four biomes.

The biomes of Africa

Yacouba Sawadogo, the man who stopped the desert

1 Read the news article below.

> ### The man who stopped the desert
>
> Yacouba Sawadogo, a peasant farmer from Burkina Faso, is famous as the 'man who stopped the desert'. Thirty years ago the land had become barren and many were giving up farming and migrating to urban areas. 'The traditional farming method used in my village allowed the rainwater to be easily washed away, leaving the crops to dry up within a short space of time. That's why I thought of a technique that would counter this problem,' said Sawadogo. His technique, called Zai, is based on traditional African farm practice. He dug holes in the soil and filled them with manure and compost. Seeds were planted in the holes at the start of the rainy season. These attracted termites that built tunnels in the hard ground, helping retain the rain. He planted trees to hold back the desert. He chose trees with medicinal properties because at the time there were no health clinics in the area. Sawadogo's community thought him mad but he persisted, and today he has a forest covering 25 hectares (62 acres).

2 Who is Yacouba Sawadogo?

_____

3 How did he stop desertification in the Sahel?

_____

_____

4 Yacouba has become world famous as a result of his approaches. A film has been made about him. Explore the website that promotes this film: www.1080films.co.uk/yacoubamovie/index.htm.

Using a computer, design a poster promoting the film. Explain who Yacouba is, the techniques he developed to stop desertification and why he has become so famous. You could find further information about him by conducting your own internet search.

Stick a printout of your poster in the space provided below, over the reminders.

Remember to include the following in your poster:

- his name
- examples of his farming techniques that held back the desert
- reasons why his techniques are working
- a title to hook people into your poster
- weblinks that you have used.

12.7 What are the challenges and opportunities of population change in Africa?

The changing population of Africa

1 Use the figures in the table below to draw a line graph.

Africa population projections	
Year	Population
2035	1,896,703,000
2040	2,100,301,000
2045	2,311,561,000
2050	2,527,556,000

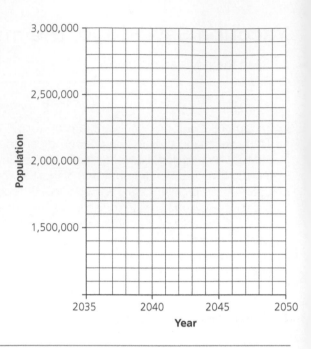

Calculate by how many people the population of Africa is predicted to grow between 2035 and 2050.

_____

2 Describe the predicted population change in Africa from 2035 to 2050.

_____

_____

_____

The future is bright for Africa's young population

Africa is young – and it continues to get younger as populations around the world are getting older. With over 40 per cent of its working age population between the ages of 15 and 24, it is the youngest continent in the world. There are almost 200 million youths in Africa and, according to African Economic Outlook, that number will double by 2045.

3 Read the quote about the population of Africa, above.

Think about what you have learnt about world economies and population and identify advantages and disadvantages of this young population.

Advantages	Disadvantages

12.8 What are the challenges and opportunities of urbanisation in Africa?

12.8

Student's Book
pages 236–237

Urbanisation of Africa

1 Think about what you have already learnt about African countries in this unit. Also consider what you learnt about urbanisation in Unit 10 to help you answer this question.

Now annotate the diagram of Lee's theory of migration to show the push and pull factors that you think are leading to the rapid growth of cities in Africa.

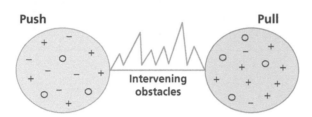

E. S. Lee's migration model

2 Go to the GiS map on the website http://luminocity3d.org/WorldCity/ and answer the following questions.

a) What does this map show?

_____

_____

b) Compare and describe the distribution of urbanisation across Africa, compared with other continents.

_____

_____

_____

c) Identify the five largest cities in Africa in rank order, from the map, and complete the table below.

Largest cities	Country	1950 Population	1990 population	2015 population	2035 population
1					
2					
3					
4					
5					

d) Look carefully at this data. When is the most rapid growth in these cities taking place?

_____

_____

12.9 Does China want to help develop Africa?

China funding the development of Africa

1 Article A from Lesson 12.9 identified one project in Africa funded by China, but there are many more. Use the internet to search for 'China funded projects in Africa'.

Write a list of five major projects in the table below.

Project	Project description – type of economic activity	African country	Cost of investment
1			
2			
3			
4			
5			

2 Read the *Guardian* online news article about China's Belt and Road policy: https://www.theguardian.com/cities/ng-interactive/2018/jul/30/what-china-belt-road-initiative-silk-road-explainer.

Use it to answer the following questions:

a) What is the policy?

_____

_____

b) Why is China investing in projects around the world as part of this policy?

_____

_____

_____

c) Why is Chinese investment in roads and railways so important to African countries?

_____

_____

_____

3 Write a paragraph to answer the lesson's enquiry question 'Does China want to help develop Africa?'

_____

_____

_____

_____

_____

12.10 What are the challenges and opportunities facing Africa? Review

12.10

Review

1 Reflect on what you have learnt about Africa in this unit, and consider the enquiry question for the unit. Summarise your understanding of the challenges and opportunities facing Africa in the table below, giving each challenge and opportunity a title.

Challenges	Opportunities

2 Label the Development Compass Rose to categorise the challenges and opportunities you identified in your table. Label challenges in blue and opportunities in black.

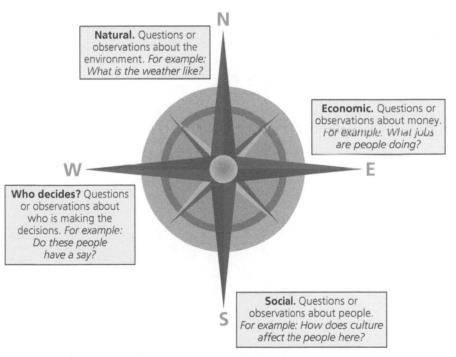

Natural. Questions or observations about the environment. *For example: What is the weather like?*

Economic. Questions or observations about money. *For example: What jobs are people doing?*

Who decides? Questions or observations about who is making the decisions. *For example: Do these people have a say?*

Social. Questions or observations about people. *For example: How does culture affect the people here?*

Definitions of development

3 Think back to Chimamanda Ngozi Adichie's comments in Lesson 12.1 about a single story view of Africa. Identify five key points that show how your story and perception of Africa has changed.

1
2
3
4
5

13.1 How does ice change the world?

The world's glaciated regions

1   Mark and name the eleven glaciated areas on the map.

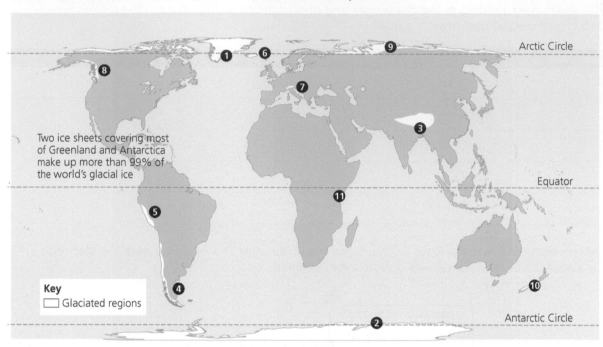

Two ice sheets covering most of Greenland and Antarctica make up more than 99% of the world's glacial ice

Key
☐ Glaciated regions

The world distribution of ice sheets and glaciers

2   Compare the map above with the map below showing the world's climate zones. Match the glaciated areas of the world to the climate zones they are in. Circle and name each glaciated area on the climate zone map.

3   Use the key from the map below to name which climate zones the glaciers are located in.

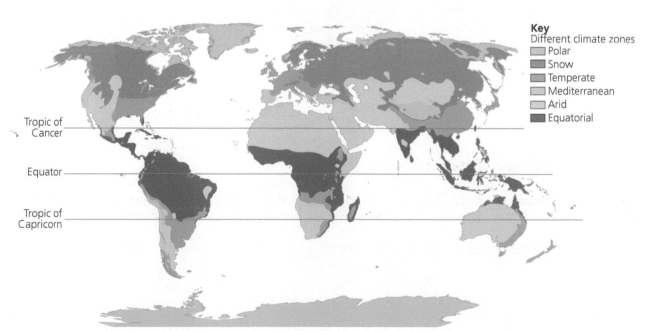

Key
Different climate zones
☐ Polar
☐ Snow
☐ Temperate
☐ Mediterranean
☐ Arid
☐ Equatorial

Climate zones

How do glaciers develop?

1 What is an ice age?

_____

_____

2 What are glacials and interglacials?

_____

_____

3 Label the glacials and interglacials on the graph below.

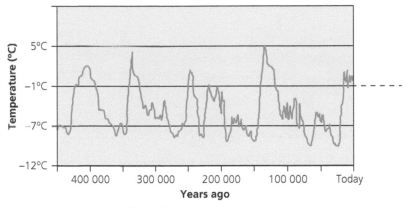

Glacial-interglacial cycles over the past 450,000 years

4 By how many degrees do world temperatures change during glacials and interglacials?

_____

A New Zealand geography teacher has made his own video on location at the country's glaciers. You will find it helpful to watch this video www.youtube.com/watch?v=VxasLA5EF9E before completing question 5.

5a Label the names of zones 1 and 2, and the main inputs and outputs, on the diagram below.

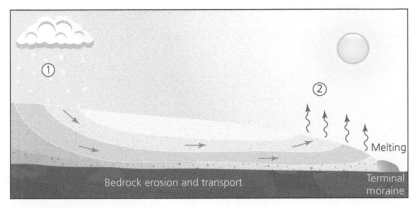

A glacier as a system

5b Annotate the diagram to explain how glaciers advance and retreat.

13.3 How do glaciers change landscapes?

Glacial erosion, transportation and deposition

1 Draw your own annotated diagrams in the boxes below to explain how glaciers erode, transport and deposit material. Make sure you use and explain the following terms when annotating your diagrams.

> plucking   abrasion   striations   zone of ablation   zone of accumulation   snout   meltwater   moraine

Glacial erosion

Glacial transportation

Glacial deposition

13.4 How are landforms shaped by glacial erosion? Part 1

13.4

Student's Book
pages 248–249

Glacial landforms formed by erosion

1   Label the geographical landforms on each diagram and annotate how landforms change.

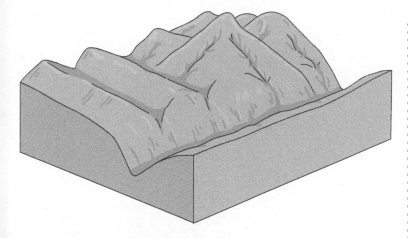

Formation of valleys: Before

Describe this landscape.

_____

_____

_____

_____

_____

_____

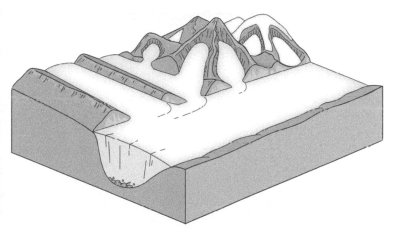

Formation of valleys: During

Describe how the ice is changing this landscape.

_____

_____

_____

_____

_____

_____

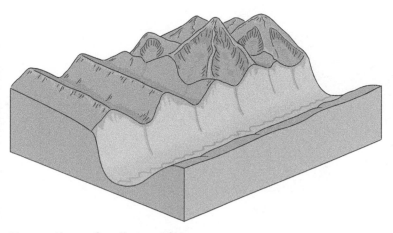

Formation of valleys: After

Describe how this landscape has changed.

_____

_____

_____

_____

_____

_____

13.5

Student's Book
pages 250–251

13.5 How are landforms shaped by glacial erosion? Part 2

Glacial landforms formed by erosion

A V-shaped valley on the road between Darcha and Rohtang Pass, Himachal Pradesh, India

A U-shaped valley in Nant Ffrancon Pass, Snowdonia

1 Compare the two photos above. Annotate clues on each photo to show which landscape was eroded by a river, and which landscape was eroded by a glacier. Label the characteristic features of each landscape.

Geirangerfjord and Seven Sisters waterfall and hanging valley

2 Label features of glacial erosion shown on the photo above.

3 Annotate these features to explain how they were formed.

Using an OS map to identify evidence of glaciation

1 Draw and shade the different glacial features shown on the OS map.

   a) Draw around the edge of corries, arêtes and pyramidal peaks.

   b) Mark and shade U-shaped valleys.

   c) Label examples of hanging valleys, truncated spurs and ribbon lakes.

2 Look carefully at the contour patterns, thinking about what you have learnt about glacial erosion in this unit. Draw arrows to show the direction you think ice moved to create Grisedale and Ullswater.

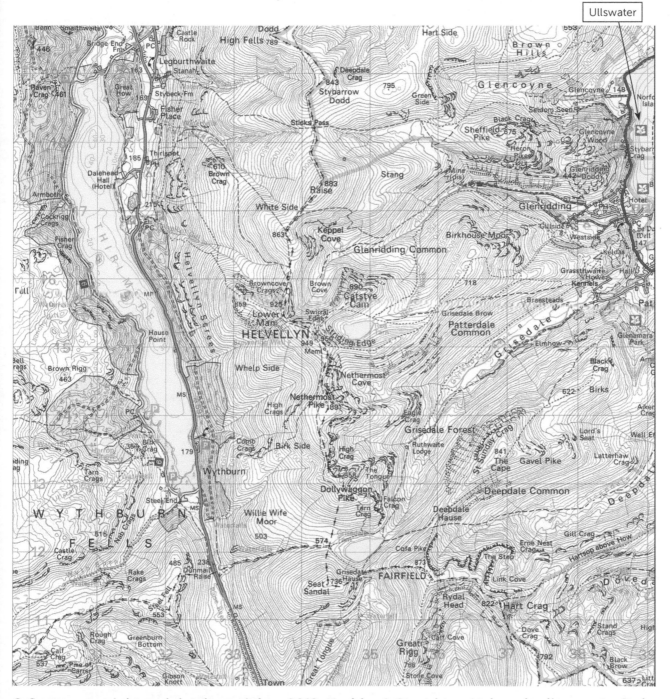

© Crown copyright and database rights, 2019, Hodder & Stoughton Ltd, under licence to Ordance Survey. License number 100036470

1:50 000 OS map extract of Helvellyn

Glacial landforms formed by deposition

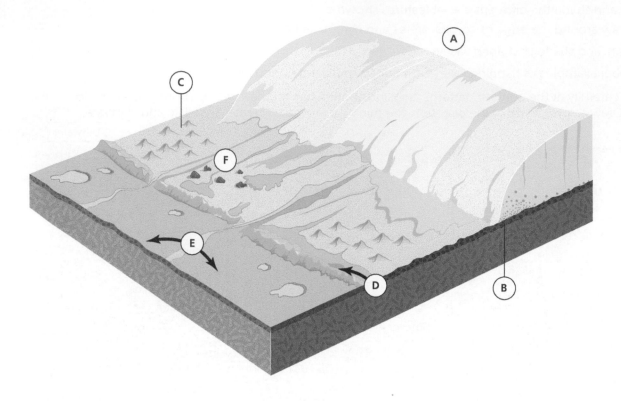

Glacial landforms formed by deposition

1   When and why does a glacier deposit material it has transported?

_____

_____

_____

2   Label the features of glacial deposition A–F on the diagram.

3   Categorise the depositional landforms into two groups in the table below.

Landforms formed from melting ice	Landforms formed from meltwater

Using an OS map and a photograph to identify how people use glacial landforms

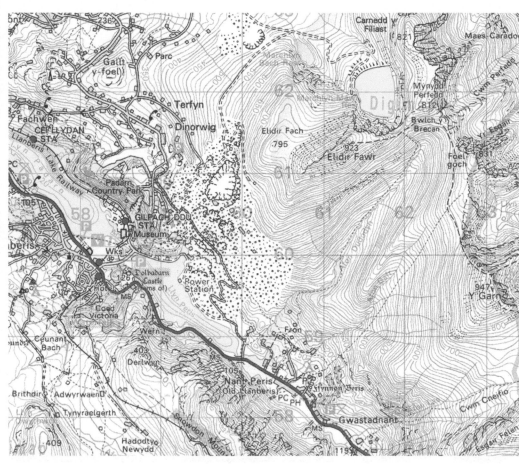

© Crown copyright and database rights, 2019, Hodder & Stoughton Ltd, under licence to Ordance Survey. License number 100036470

1:50 000 OS map extract of Dinorwig

1 Draw around the contour patterns on the OS map to show glacial landforms. Label the landforms on the map.

2 Label evidence on the map of how people in the area are using glacial landforms.

3 Annotate the map to explain how the landforms are used.

4 Look carefully at the photo of Geiranger in Norway.

   a) Label the following on the photo: fjord, village, cruise ships, campsite, hotels.

   b) Annotate the photo to describe what this place is like using the five enquiry questions. You may find it helpful to visit the Geiranger port website: www.stranda-hamnevesen.no

Geiranger

13.9 How do we investigate how glaciers are changing?

The work of glaciologists

1 Watch the following two video clips showing the work of glaciologists.
www.youtube.com/watch?v=NdpJ4L5qMy8
https://vimeo.com/18713375

2 How and why do glaciologists investigate glaciers?

_____

_____

_____

3 Investigate the following websites.

> USGS Repeat photography project:
> www.usgs.gov/centers/norock/science/repeat-photography-project?
> qt-science_center_objects=0#qt-science_center_objects
>
> Mendenhall glacier time lapse using remote cameras
> https://vimeo.com/2637977
>
> Using satellite image remote sensing to study glaciers:
> www.antarcticglaciers.org/glaciers-and-climate/glacier-recession/
> observing-glacier-change-space/

4 Using the evidence you researched on the websites, complete the table to explain some of the techniques glaciologists use to investigate glaciers.

Technique	Explanation
Ice cores	
Repeat photographs	
Remote camera	
Satellite images	

5 Use the evidence from the websites to write four sentences to explain what you think glaciologists have discovered about how glaciers are changing.

_____

_____

_____

_____

_____

13.10 How does ice change the world? Review

13.10

Student's Book
pages 260-261

Review

1 Label the glacial landforms 1–10 shown on the diagram.

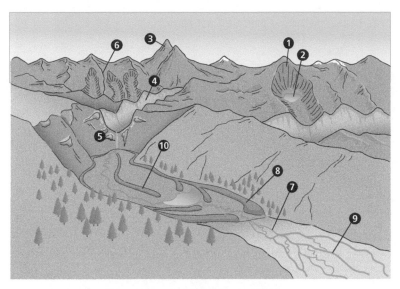

Features associated with alpine glaciation

2 The mind map below has been started to help you reflect on your learning for this unit. Complete the mind map – you can add more strands, simple drawings, and different colours to more clearly identify links.

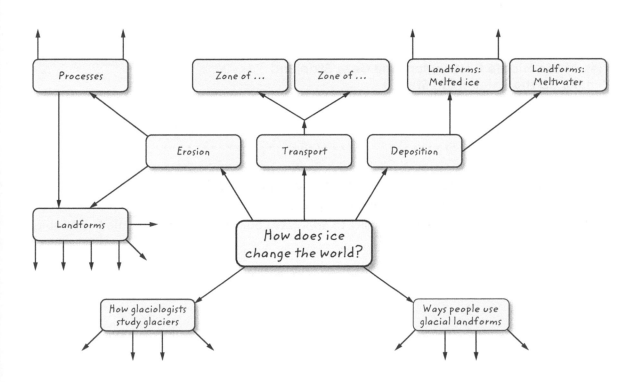

14.1 Why is the Middle East an important world region?

What is the region of the Middle East like?

1 Using an atlas, find and label the following on the map below:

- the countries of the region
- the three continents that border the Middle East region.

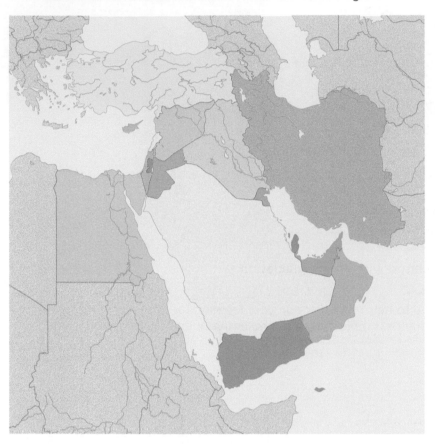

The countries of the Middle East region (the Gaza Strip and West Bank make up the Palestinian States)

2 Go to the degree confluence website www.confluence.org. Search for a confluence point in the region. Download and print a photo of a location in the Middle East and stick it in the space below.

Things to do with the photo

- Add a title naming the place and providing its location using latitude and longitude.
- Mark and name the location of the place on the map above.
- Use the description of the place on the website to annotate your photo, answering the five enquiry questions.

14.2 How does physical geography influence the region?

Research the physical geography of the Middle East

1 Compare the map below with an atlas map of the Middle East.

   a) Mark and name the seas that surround the continent.

   b) Mark and name what you think are the major mountains, deserts, lakes and basins.

   c) Mark and name the following major rivers: the Tigris, Nile and Euphrates.

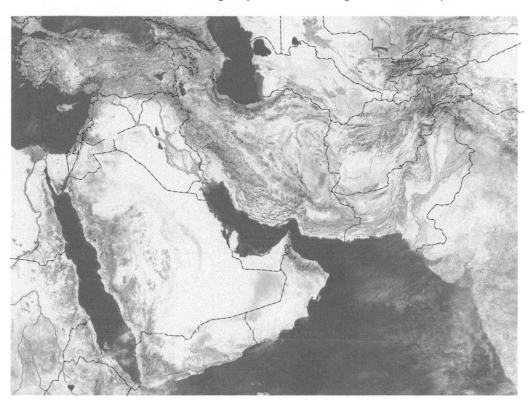

The Middle East physical geography

2 Research the following website: www.geolsoc.org.uk/Plate-Tectonics/Chap3-Plate-Margins/Divergent/Triple-Junction. Using your research, mark on the map where you think volcanoes and earthquakes are likely to occur in the Middle East.

   Write four sentences to explain why volcanoes and earthquakes occur in the Middle East.

_____

_____

_____

_____

_____

_____

_____

_____

14.3 What problems does the climate of the Middle East create for the region?

The climate of the Middle East

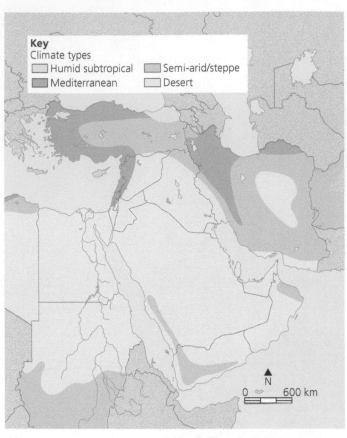

Key
Climate types
☐ Humid subtropical   ☐ Semi-arid/steppe
☐ Mediterranean   ☐ Desert

N
0   600 km

Climate types in the Middle East

1 Annotate the map to describe the different types of climate across the Middle East.

2 Why do you think water is a precious natural resource in the Middle East?

_____

_____

_____

_____

_____

_____

3 Read viewpoints A and B and explain how they are improving water supply in the region.

A Farming in arid areas traditionally uses flood irrigation, which wastes a lot of water. In Israel, farmers use recycled water and have also developed drip irrigation – small plastic pipes in fields that put water right at the roots of crops. This saves 40 per cent of water used in farming.

B In Saudi Arabia, 40 per cent of our water comes from underground aquifers, where the rock is saturated with water stored for thousands of years. We need to be more careful that we don't use it too quickly, as it is non-renewable.

_____

_____

_____

_____

4 Explain which approach you think is conserving water use.

_____

_____

14.4 Why is the population of the Middle East so diverse?

The population distribution of the Middle East

1   The map below shows the population density across the Middle East. Look carefully at the key and decide which colours of shading represent densely populated, and which represent sparsely populated areas. Add labels to the key to show this.

2   Now annotate the map to show densely and sparsely populated areas.

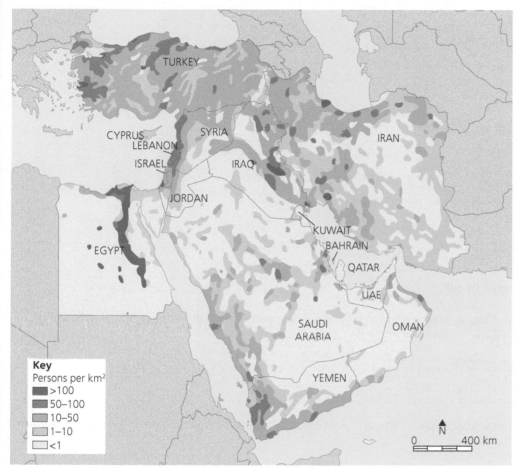

Key
Persons per km²
>100
50–100
10–50
1–10
<1

Population density across the Middle East region

3   Think about what you have learnt so far about the Middle East, and use an atlas to consider the reasons for the population distribution across the region. In the table below, explain how physical geography and climate influence this distribution.

Factor	Sparsely populated areas	Densely populated areas
Physical geography – coast, mountains, flatland, rivers		
Climate		

14.5

14.5 Why is the Middle East a major economic region of the world?

Student's Book pages 270-271

The global balance of trade in oil

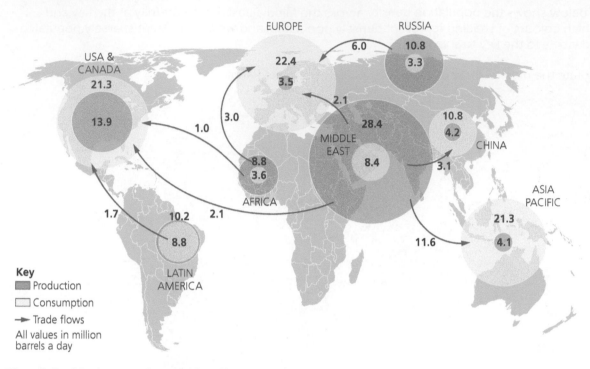

Key
- Production
- Consumption
- → Trade flows

All values in million barrels a day

The global balance of trade in oil

1 What do the Middle East, Russia, Africa, and Latin America have in common, in terms of oil production and consumption?

_____

_____

2 What do the USA and Canada, Europe, China and Asia Pacific have in common in terms of oil production and consumption?

_____

_____

3 Which of the two groups of regions in Questions 1 and 2 are in the strongest position in terms of the global balance of oil? Explain your answer.

_____

_____

4 The Middle East is in the strongest position; explain why.

_____

_____

Levels of development in the Middle East

1 The world map below shows the world distribution of GNI per capita. Compare it with your map of the Middle East in 14.1. Identify the location of each country in the Middle East on the world map below and write the names of each country in the Middle East around the map in four GNI per capita income groups.

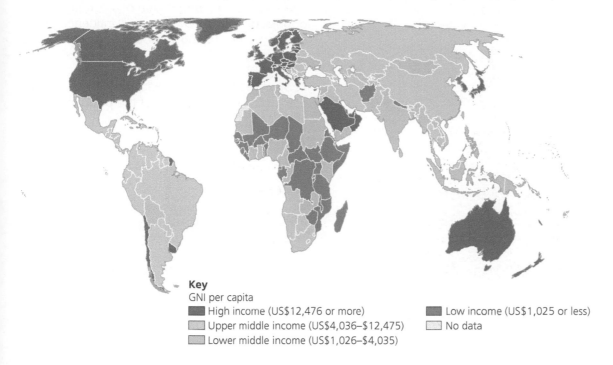

Key
GNI per capita
■ High income (US$12,476 or more)   ■ Low income (US$1,025 or less)
■ Upper middle income (US$4,036–$12,475)   ▢ No data
■ Lower middle income (US$1,026–$4,035)

Global distribution of GNI per capita, 2017

2 Describe the distribution of development in the region.

_____

_____

3 The United Arab Emirates has used income from oil to diversify its economy, reduce its independence on oil exports and therefore develop. Use the following websites to identify two ways the country has diversified.

Give one reason why it was important that the UAE reduced its dependence on oil.

www.uae-embassy.org/news-media/uae-economic-diversification-efforts-continue-thrive
https://vimeo.com/76076906

_____

_____

_____

_____

_____

14.7 Why is Yemen the poorest country in the Middle East?

Population change in Yemen

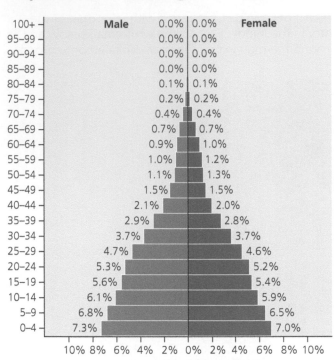

Population pyramid for Yemen 2017 (total population of 28,119,545)

1 The population of a country can be divided into three age groups – young, aged 0–15, working or economically active, aged 16–65, and old age, over 65. The proportion of a country's population in each category will influence how a country is developing, as you discovered in Units 8, 10 and 12.

Look at the population pyramid for Yemen and calculate the percentage of the population in each age group.

Young: 0–15 _____%

Economically active: 16–65 _____%

Old 65+: _____%

(Hint: remember to add together both the male and female figures for each age group.)

2 Write two sentences about what these percentages suggest is happening to the population in Yemen.

_____

_____

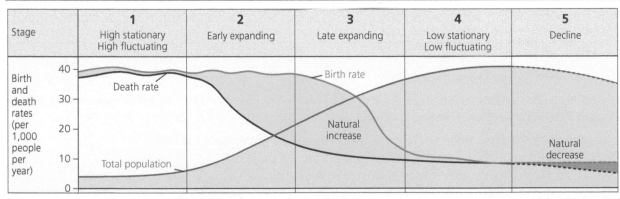

The Demographic Transition Model

3 Look carefully at the Demographic Transition Model above and compare it with the population pyramid for Yemen to decide which stage of the model you think Yemen's population is in. Annotate the evidence on the model that helps explain your choice.

4 Using the evidence from the pyramid and the model, write two sentences to explain what you think is happening to the population of Yemen, and how this contributes to Yemen being a poor country.

_____

_____

_____

Reasons for conflict in the Middle East

1 Complete the sentences using the words from the box.

| colonies | Shia Muslim | oil | Israel | Arab Spring | USA | Sunni Muslims | Palestine |

Wars occur on a regular basis in the Middle East, for a variety of reasons. Many country borders in the region were created artificially by Britain and France, when the regions were _____. _____, for example, was created in 1948 as a Jewish homeland, dividing _____, causing ongoing conflict. In 2011 protesters across the Middle East region took to the streets, demonstrating against their governments, where unemployment and corruption were common. This was called the _____ and led to a change of government in several countries and ongoing armed conflict in others. _____ and the wealth generated by it is a significant cause of conflict in the region. It has led to the interference by countries like the _____ in conflict, to protect their economic interest, which only intensifies conflict. There is religious division in the region between two Islamic sects, the _____ and the _____.

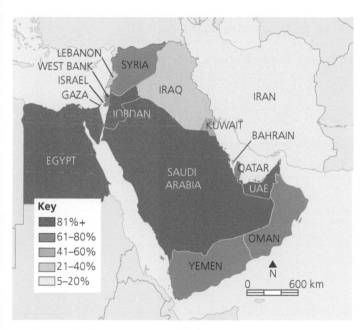

Estimated distribution of Sunni Muslims in the Middle East

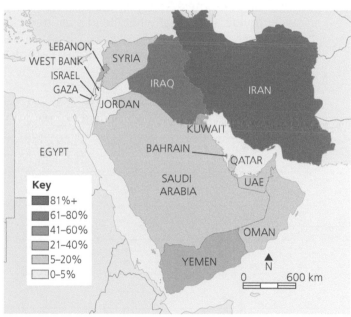

Estimated distribution of Shia Muslims in the Middle East

2 Write two lists naming the countries where either Sunni Muslims or Shia Muslims form the highest percentage of the population.

_____

_____

_____

3 Annotate the two maps to explain why you think that Saudi Arabia is fighting with the Sunni Muslim government of the Yemen against the Shia Houthi rebel group, and why Iran opposes military action in the Yemen.

14.9

Student's Book
pages 278-279

14.9 Why is the Middle East an important world region? Part 1

The refugees of Syria

1 Visit this website www.bbc.co.uk/news/world-middle-east-32057601. The site has a role play decision-making activity that puts you in the shoes of a Syrian refugee.

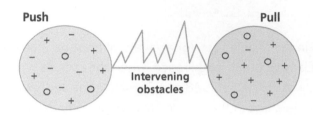

Using your research of the BBC role play website, annotate Lee's migration model to summarise why you think Syrian refugees make the decision to flee the conflict in their country.

Identify the push and pull factors and the obstacles they have to overcome in their forced migration.

2 Visit the interactive Refugee Republic website https://refugeerepublic.submarinechannel.com. This site will provide an idea of what life in a refugee camp is like for Syrians fleeing the conflict.

Describe what life is like in the refugee camps and why you think refugees often decide to move again from the camps.

_____

_____

_____

_____

_____

_____

_____

_____

_____

_____

_____

_____

_____

_____

14.10 Why is the Middle East an important world region? Part 2

14.10

Student's Book
pages 280–281

Review

1 Reflect on what you have learnt in this unit of work about the Middle East, and summarise how you have progressed your understanding of the big ideas of geography shown in the boxes below.

(Hint: give examples from your studies of the Middle East.)

The Middle East as a region

_____

_____

_____

_____

Population

_____

_____

_____

_____

Plate tectonics

_____

_____

_____

_____

Migration

_____

_____

_____

_____

Economy

_____

_____

_____

_____

2 Conclusion: Identify five things that make the Middle East an important world region.

1 _____

2 _____

3 _____

4 _____

5 _____

15.1 What is the future for the planet? A geographer's view

Different views about climate change

1 Read the following points of view about climate change.

> There is only 12 years for global warming to be kept to a maximum of 1.5°C, beyond which even half a degree will significantly worsen the risks of drought, floods, extreme heat and poverty for hundreds of millions of people. We have presented governments with pretty hard choices. We have pointed out the enormous benefits of keeping to 1.5°C. We show it can be done within laws of physics and chemistry. Then the final tick box is political will. We cannot answer that. Only our audience can – and that is the governments that receive it.

Intergovernmental Panel on Climate Change (IPCC) scientists' report, 2018

> I think something's happening. Something's changing and it'll change back again. I don't think it's a hoax. I think there's probably a difference. But I don't know that it's manmade. I will say this: I don't want to give trillions and trillions of dollars. I don't want to lose millions and millions of jobs.

> Climate change brings in more favourable conditions and improves the economic potential of this region, allowing us to better extract oil and minerals, as well as possible new northern shipping lanes as the ice recedes.

President Trump's response to the IPCC report, 2018   **Vladimir Putin, Russian president, 2017**

2 What issues have the panel of scientists presented for world governments in their 2018 report?

_____

_____

3 What do these scientists consider their role to be in achieving their goals, and whose job is it to act upon them?

_____

_____

4 Explain how President Trump responded to the report.

_____

_____

5 What is President Putin's view of climate change?

_____

_____

_____

6 Why do these views of world leaders make climate change controversial and difficult to resolve?

_____

_____

_____

15.2 What is the evidence for climate change?

Analysing evidence of climate change

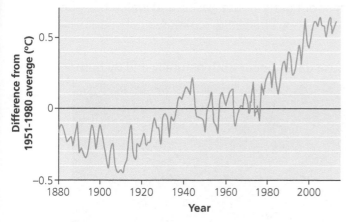

Graph to show changes in annual global temperatures, 1880–2013

1 Annotate the graph to show how the data provides evidence of climate change.

2 Write three sentences to explain how the world's climate is changing.

_____

_____

_____

_____

_____

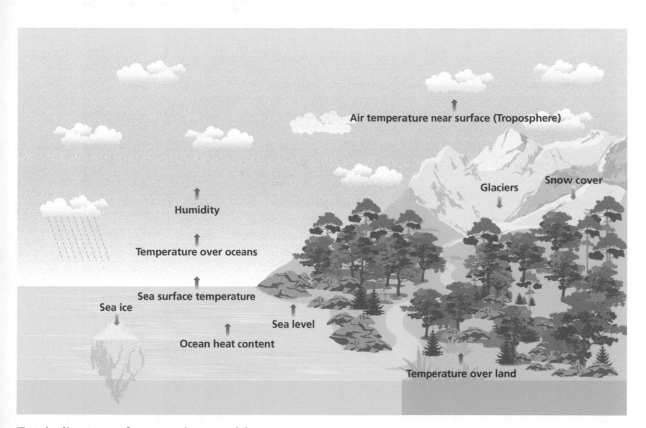

Ten indicators of a warming world

3 Annotate the diagram to explain how changes in these ten indicators give evidence of climate change and are interconnected; for example an increase in temperature leads to a decrease in size of glaciers.

15.3 What are the causes of climate change?

The greenhouse effect

1 Go to the following website, scroll down to find out about greenhouse gases, and watch the video explaining the greenhouse effect: www.activesustainability.com/climate-change/what-is-the-greenhouse-effect/

2 Annotate the diagram below to show how greenhouse gases occur both naturally and by human action.

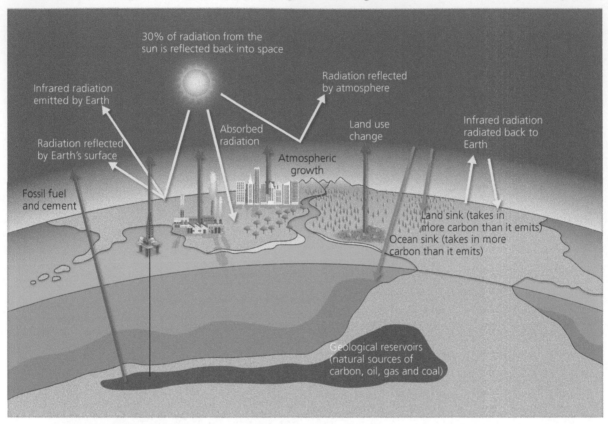

The greenhouse effect

3 Explain how increases in greenhouse gases can lead to climate change.

_____

_____

_____

_____

_____

_____

_____

_____

_____

Hodder & Stoughton Limited © David Gardner 2019

15.4 What are the consequences of climate change for our planet? Part 1

15.4

Student's Book
pages 288–289

Researching the consequences of climate change

1 Make a list of five major consequences of climate change in the table below. Research the consequences of climate change using the internet. The following three websites will make a good start.

https://vimeo.com/79771046
www.bbc.co.uk/news/science-environment-24021772
www.bbc.co.uk/news/entertainment-arts-47988337

You can widen your research using a search engine such as Google.

	Consequence	Description of impact on people and the planet
1		
2		
3		
4		
5		

2 Many of the consequences of climate change are interconnected. Think about what you have already learnt about how the Earth's spheres interact and are interdependent. Go back to lessons 13.9 and 15.2 in your workbook and consider the consequences of climate change on glaciers, and about how melting glaciers and ice sheets have further consequences. Complete the flow diagram below to identify how one change to the Earth's spheres has consequences for the others.

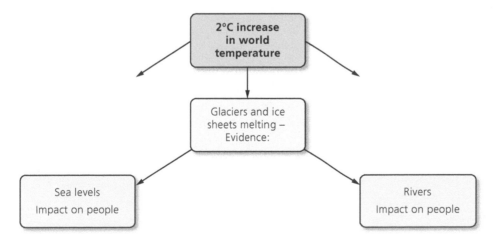

15.5

15.5 What are the consequences of climate change for our planet? Part 2

Student's Book pages 290–291

Create a campaign poster supporting people affected by climate change

ActionAid and Oxfam have developed climate change campaigns because they believe people living in poverty are most vulnerable to its impact.

Imagine you are a graphic designer employed by one of these organisations to produce a campaign poster explaining this impact. You can create your poster using a computer and stick a print-out of it in the space below.

Research their campaigns at the following websites to develop your evidence and ideas for the poster.

https://actionaid.org/land-and-climate

www.oxfam.org.uk/what-we-do/issues-we-work-on/climate-change

Stick your poster here.

Remember to:
- include the name and logo of the organisation
- include a catchy title to hook people in to look at your poster in detail
- use photos and data to provide evidence of the impact of climate change on poverty
- highlight key facts.

15.6 What are the consequences of climate change for the UK?

15.6

Student's Book pages 292–293

Impact of climate change on the UK

1 The UK Government has provided guidance for the public about climate change:
www.gov.uk/guidance/climate-change-explained#climate-change-now, as well as a video
www.youtube.com/watch?v=SDxmlvGiV9k

Use this guidance to summarise in the table below how the government believes climate change
will impact on the UK.

Impact of climate change	Effects
Rising temperatures	
Rainfall patterns and water supply	
Oceans	
Food production	
Ecosystems	
Human health	

2 Read the quote from Sir James Bevan, the Chief Executive of the Environment Agency.

Highlight in red any text in the quote about the consequences of climate change and use yellow to
highlight what he states the Environment Agency is doing to protect people in the UK. Highlight in green
what he suggests people need to do themselves.

Climate change is likely to mean more frequent and intense flooding. Floods destroy – lives, livelihoods, and property. Our flood defences reduce the risk of flooding, and our flood warnings help keep communities safe when it threatens. But we can never entirely eliminate the risk of flooding. This is why we have launched guidance on flooding. Checking your flood risk is the first step to protecting yourself, your loved ones and your home.

Sir James Bevan

3 Go to the Environment Agency website to check the long-term risk long term of flooding for your home:
www.gov.uk/check-flood-risk

Enter your postcode to identify the long term risk of flooding and summarise your findings:

at your home _____

at your school _____

in your local town _____

15.7

15.7 Antarctica – the frozen continent? A geographical enquiry

Student's Book
pages 294–295

Conducting your initial research about Antarctica

Use an atlas, Google Earth and websites to discover important information about Antarctica.

Record any key information in the table below.

Useful websites

https://discoveringantarctica.org.uk

www.bas.ac.uk/

www.asoc.org/index.php

www.antarctica.gov.au/webcams

www.coolantarctica.com/index.php

Key information	Key facts and data
Where is it?	
What is it like?	
Climate	
Landscape	
How is Antarctica changing as a result of climate change?	
Reasons to protect Antarctica	

The USA withdrawal from the Paris Agreement

1   Watch the WWF video about the Paris Agreement at www.youtube.com/watch?v=I-4F5MJEeqs.

Explain what the agreement is trying to achieve.

_____

_____

_____

_____

_____

2   In June 2017 the USA withdrew from the Paris Agreement.

Watch the two video clips of President Trump's speech announcing USA withdrawal and President Macron of France's reaction:

www.bbc.co.uk/news/av/world-us-canada-40127876/trump-explains-why-us-pulling-out-of-paris-accord

www.bbc.co.uk/news/av/world-europe-40125579/paris-climate-deal-macron-pledges-to-make-planet-great-again

In the speech bubbles below summarise the different views of these two presidents.

President Donald Trump	President Emmanuel Macron

3   Think about what you have learnt so far in this unit. Explain the problems you think the US withdrawal will create for meeting the targets of the Paris Agreement, and for the future of the planet.

_____

_____

_____

_____

_____

15.9

15.9 What is the future for the planet? A geographer's view: Review

Student's Book
pages 298–299

Create a concept map to show the big ideas of geography and how they link

Make interconnections between the big ideas of geography you have studied in the Progress in Geography course, and how they affect climate change.

Annotate the connection lines to explain how each big idea interconnects.

Draw red lines for the causes of climate change and blue lines for the effects.

List the causes and effects of climate change for each big idea in each box.

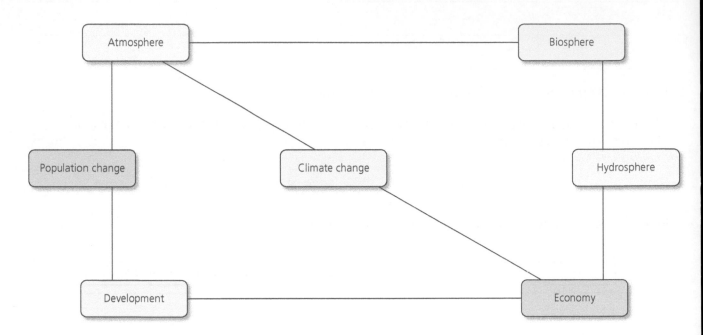

The big ideas of geography and how they link to climate change

Progress to GCSE Geography

Progress in Geography has been designed to help you become a good geographer, and provide the key skills, knowledge and understanding for you to build on in order to progress and succeed at GCSE level.

Have a look at the GCSE specification for Geography that is taught at your school.

1 Complete the table below to explore the content of the GCSE course and how your learning through *Progress in Geography* helps prepare you for your future studies.

What you will study at GCSE	How your learning from *Progress in Geography* helps

2 Write a summary paragraph to explain how *Progress in Geography* has prepared you for the GCSE Geography course.

_____

_____

_____

_____

PROGRESS IN

WORKBOOK 3

GEOGRAPHY

UNITS 11–15

KEY STAGE 3

Review and reinforce the skills, knowledge and understanding that you are developing throughout your Progress in Geography: Key Stage 3 course.

This Workbook accompanies your Student Book, providing extra support as you continue on your journey to become a good geographer.

- A range of activities focus on skills, knowledge and understanding
- Ideal for homework, classwork and independent study
- One Workbook page for every lesson in the Student Book

Also available:

Workbook 1: Units 1–5 (Single copy)
ISBN: 9781510428072

Workbook 1: Units 1–5 (Pack of 10)
ISBN: 9781510442993

Workbook 2: Units 6–10 (Single copy)
ISBN: 9781510428065

Workbook 2: Units 6–10 (Pack of 10)
ISBN: 9781510443006

Workbook 3: Units 11–15 (Pack of 10)
ISBN: 9781510443013

With special thanks to

The world's trusted geospatial partner

HODDER EDUCATION
t: 01235 827827
e: education@hachette.co.uk
w: hoddereducation.co.uk

ISBN 978-1-5104-4298-6

9 781510 442986

MIX
Paper from
responsible sources
FSC™ C104740